# THE FISH
## WHO TOUCHED
## THE LIGHT

*The Story of the Viral Anglerfish*

## Felix Grayson

**MINDSPARK**
PUBLISHING

*To the ones who rise.*
*To the dreamers, the wanderers, the souls who
chase the impossible.*
*To those who leave behind the familiar, reaching
for something greater —**even when they don't
understand why.***
*This is for you.*
*May you always have the courage to touch the
light.*

"You cannot swim for new horizons until you have courage to lose sight of the shore."

— William Faulkner

# CONTENTS

# PROLOGUE: A LIGHT IN THE DARKNESS

There is no light here.

Not a flicker. Not a shadow. Not even the memory of something that once glowed and then faded.

The deep is not just dark—it is **absence itself.**

Here, a thousand feet beneath the waves, the sun has no reach. No colors exist, no warmth lingers. It is a world of silence, untouched by time, ruled by pressure and patience.

And within it, she drifts.

A creature of shadow, of stillness, of quiet survival.

She does not know of light, not beyond the one she carries—the tiny glow at the end of her lure, the small, flickering promise she uses to bring life toward her. It is not for warmth. It is not for beauty.

It is **a trap.**

Because here, in the deep, light is not a gift. It is a

trick. A whisper of hope in an ocean that has none.

And yet—something stirs.

A pull. A shift in the water. A force she cannot see, but feels deep within her.

It is not hunger. It is not danger.

It is something else.

Something **unknown.**

She does not understand it. But she obeys.

Her fins, so used to drifting, begin to move.

And for the first time, she is not waiting.

She is not hunting.

She is **rising.**

# CHAPTER 1: **THE DEEP, UNKNOWN WORLD**

## THE DEEP, UNKNOWN WORLD

Beneath the restless waves of the ocean, beyond the golden sunlight that dances upon the surface, past even the dim blue twilight where sharks and wandering giants glide—there is a world of silence. A world unseen. A world untouched by light.

It is called the **hadal zone**, the abyssal plains, the midnight depths—where the weight of the ocean above is so immense it would crush a human in an instant. The water here is black, eternal, and absolute. If the deep sea were a planet, it would be one of endless night, a place where the concept of sky and stars is meaningless, where the idea of light itself is a distant fantasy.

And yet, it is not empty.

Here, creatures exist that defy reason. They drift like specters, their bodies shaped by an environment so extreme that life should not be possible. There are fish with transparent heads, others with mouths so wide they can swallow prey their own size. Some pulse with eerie blue glows, while others stretch delicate filaments like ghostly fishing lines into the void, waiting for the unsuspecting to wander too close.

This is where **she** was born.

She has never known warmth. Never felt the caress of current against coral, never seen the sky above, never known color beyond the pale glimmer of her own tiny lure. To her, the world is not a place of landscapes but of drifting shapes and unseen currents. She is a phantom in an ocean of shadows, her body designed for darkness, her existence written in the language of patience and hunger.

The pressure here is **bone-crushing, a thousand times greater than at the surface.** The water is still and cold, a steady **one to four degrees Celsius**, unchanging, indifferent. No seasons exist here, no tides to mark the passing of time. There is only the abyss, stretching endlessly in all directions.

She has never questioned her place. She has never needed to. The deep is all she has ever known. The deep is all she was ever meant to know.

And yet… something calls.

A force she does not understand. A pull, an instinct, an urge buried so deep within her that even she does not know why she obeys.

And so, she moves.

Not forward. Not sideways.

But **upward.**

## A PLACE OF SILENCE, PRESSURE, AND CREATURES OF NIGHTMARES

Silence.

Not the quiet hum of a city at dawn, nor the stillness of an empty room. This is a silence so complete it becomes a presence of its own—a vast, unbroken emptiness stretching for miles in every direction. No birds sing here. No waves break against a shore. No wind whispers across the endless plains of black silt.

The silence is absolute.

The deep sea is a world built on pressure—an unseen force pressing against everything that dares to exist here. The deeper one travels, the heavier the ocean becomes, layering itself over every living thing like an invisible weight, a crushing hand that will not relent. Down here, in the abyss, the pressure is thousands of pounds per square inch, so intense that if a human were to descend unprotected, their body would collapse before they could even comprehend what had happened.

Yet, life does not surrender.

It twists and adapts, molding itself to the impossible.

The creatures that survive in the deep are not like those found in sunlit waters. They are **strange, spectral things** — made of translucent flesh, stretched limbs, and eyes that glow like dying embers. They are born in a place where food is a rarity, where every movement must be calculated, every hunt a test of patience.

Some have evolved to see in the dark, their eyes enormous, capturing the faintest glimmers of bioluminescent light. Others have no eyes at all, for what use is vision in a world where light has never existed? Instead, they **taste the currents, feel the pressure shifts, and move like ghosts through liquid blackness.**

And then there are the nightmares.

Creatures with jaws that unhinge to swallow prey whole. Bodies lined with needle-sharp teeth, designed not for killing, but for ensuring that anything caught can never escape. Some dangle glowing lures from their heads, tricking the foolish into coming closer, before a mouth snaps shut in an instant. Others remain motionless for weeks, waiting in the darkness for something—**anything**—to make the fatal mistake of swimming too near.

This is the world she has always known.

She is not its master. She is not its ruler. She is merely a part of its endless cycle—hunting and waiting, **a flickering light in the dark.**

And yet, something in her stirs.

Something urges her to move.

To leave behind the silence. To rise toward something unseen.

To go where no anglerfish has ever gone before.

# SOLITUDE, SURVIVAL, AND THE QUIET RHYTHM OF THE ABYSS

She drifts.

Suspended in the vast emptiness, she is a shadow among shadows, a ghost in a world where no one is watching. There is no day, no night. Time is measured not by the rising of a sun—there is no sun here—but by the slow churn of hunger in her belly and the silent flicker of her lure.

This is her world. A world of stillness, of waiting.

There is no companionship in the abyss. No family, no school to swim alongside. She has never known the touch of another except for the tiny, parasitic male that once found her—small and fragile, his only purpose to merge with her flesh, to become a part of her, nothing more than a vessel for the next generation. She carries him now, fused into her side, his body an afterthought, his existence reduced to the simple act of ensuring life continues.

But beyond that, she is alone.

Yet solitude is not loneliness. It is not sadness. It is

simply **what is.**

She does not question it, does not fight against it. She moves through the abyss with the quiet patience of something that has never known another way. She is built for this life—her body small, yet perfectly adapted to the crushing depths, her mouth a cavernous trap designed to consume whatever fate delivers to her. She does not chase, does not fight. She waits.

And when the moment comes, she is ready.

Her lure glows—a faint, pulsing beacon in the blackness. It is the only light for miles. She dangles it just so, letting it drift in the unseen currents, a whisper of movement in a place where movement is rare. She does not know what will come. A tiny, unsuspecting fish? A foolish, wandering shrimp? Or perhaps nothing at all?

She does not control the outcome. She only **waits.**

Then—a shift. A ripple in the water.

Her jaws snap open before thought can form, before instinct can register. Her teeth, curved inward, ensure there is no escape. The moment her mouth closes, it is done.

For now, she is satisfied. The hunger retreats. The waiting begins again.

This is her rhythm.

This is the abyss.

And for all the years of her life, she has never questioned it.

Until now.

Because something has changed. Something deep within her, something she cannot name.

A pull. A whisper. A feeling unlike any she has ever known.

She is moving.

And this time, she is not drifting.

This time, she is **rising.**

## THE LURE, THE HUNT, AND THE BOND THAT CANNOT BE BROKEN

In the abyss, the hunt is not a chase. It is not a battle of speed or strength.

It is patience.

It is stillness.

It is the art of waiting.

She does not swim through the darkness in search of prey. To move is to waste energy, and energy is scarce in this place where food is a fleeting miracle. Instead, she becomes **the ocean itself**, motionless, blending into the void, her body dissolving into the nothingness around her. She does not exist—only her lure does.

A tiny thread extends from her forehead, a delicate filament of flesh that twists and flickers like the light of a dying star. At the tip, a small orb glows with an otherworldly radiance, pulsing in the blackness like a beacon calling lost souls home.

This is her gift. Her secret weapon. Her deception.

Deep within her, bacteria thrive in perfect symbiosis, their bioluminescent dance the only light she has ever known. They do not belong to her, yet without them, she is nothing. They are the glow within her darkness, the heartbeat of her existence. They **illuminate the abyss, drawing life toward her—toward their end.**

The unsuspecting come.

A flicker in the distance. The ripple of a body that does not yet realize its fate. A fish, a wandering shrimp, an unknowing traveler in the endless void. They see the glow, the soft promise of light where none should be.

In the deep, **light is life**—and in their desperation, they believe.

They come closer.

She does not move. She does not breathe. She lets them believe.

And then—

**The ocean erupts.**

Her mouth is a cavern, her jaws unhinging in an instant. The water rushes in, pulling her prey forward before it can even register its mistake. There is no time to struggle. No time to flee.

Her teeth, long and curved inward, are not designed to kill. They are designed to hold. To trap. To ensure that once something has entered her grasp, it will never leave.

The darkness swallows it whole.

She blinks. Her lure flickers once more. The waiting begins again.

**But the hunt is only half her story.**

She is not alone—not entirely.

Far smaller than she, barely more than a whisper of flesh, there is another. He has no lure. No teeth. No light of his own. In the abyss, he would not last a day without her.

He is a male.

And he exists for one purpose.

He found her once, a long time ago, when he was still whole. He latched onto her, his body pressing against hers in the dark, his tiny mouth finding her flesh. And then… he began to disappear.

Slowly, he became part of her. His body, his bones,

his skin—everything but the essential—**fused into her own.** His blood merged with hers. His self became hers. He no longer hunted. No longer fed. No longer moved.

Yet, he was not dead.

He was **inside her, within her, forever a part of her.**

He would never leave. Never seek another. Never touch the world again.

And when the time came, when nature called, and the deep demanded life continue, she would take what was needed from him. He would give what he was made to give.

And the cycle would continue.

She has never questioned it.

But tonight, as she drifts in the blackness, **something within her stirs.**

Something foreign.

Something that does not belong in the abyss.

A pull. A whisper. A call from above.

And for the first time in her life... she moves toward it.

## BUT FOR REASONS UNKNOWN... SHE MOVES UPWARD

Something is wrong.

Or perhaps, something is finally right.

She does not know.

For all her life, she has been still. Suspended in the void, drifting as the abyss dictated. She has never fought against the pull of the deep, never questioned the silent current that cradled her.

Until now.

Now, there is a call—faint, unfamiliar, but undeniable.

It is not hunger. That would be familiar. It is not danger. That would be known.

This is something else.

Something ancient. Something buried deep within her flesh, her bones, her blood. A whisper from a place she has never been. A feeling that is neither instinct nor reason, but something older than both.

A pull.

And so, she moves.

It is subtle at first—a shift, a slow ascent. The water feels strange against her skin, thinner, looser, as if something is slipping away. The pressure, once her constant companion, begins to ease.

For the first time, she senses change.

She does not know that she is rising.

She does not know that this journey is not meant to be.

Her body is designed for darkness, for the relentless weight of the ocean crushing down upon her. She does not belong anywhere but the abyss. She does not understand what awaits above, what waits beyond the blackness.

But still, she moves.

She has never seen light beyond her own. Never known a world beyond this one.

Yet, something calls.

And without knowing why, without knowing how—

**She obeys.**

# CHAPTER 2: **THE ASCENT**

## RISING FROM THE DEEP IS UNNATURAL FOR HER KIND

She does not belong here.

The moment she begins to ascend, the ocean changes around her. It feels different, unnatural, as if she has crossed an invisible boundary she was never meant to breach.

The abyss has always held her, its embrace unrelenting, its weight pressing into every inch of her body. It was all she had ever known. She was made for the deep, designed for its silence, its stillness, its hunger.

And yet, she is rising.

The darkness thins. Not by much, not yet, but enough for her to sense that something is different. The water moves in ways it never has before, swirling, shifting, less like a great stillness and more like something alive.

For the first time, **she feels the ocean instead of simply existing within it.**

She does not know that she is leaving behind a place where few have ever gone and fewer still have returned.

She does not know that what is waiting above is not for her.

She only knows that something is wrong.

Not in a way that warns of danger. Not in a way that tells her to stop.

But in a way that tells her **she is going where no anglerfish has ever gone before.**

And still… she does not stop.

She does not yet feel pain. She does not yet feel the unraveling of everything she is, everything she was made to be. That will come later.

For now, there is only movement.

For now, there is only the unknown.

And still… she keeps rising.

# THE LIGHT ABOVE IS CALLING, BUT SHE DOES NOT KNOW WHY

There is something ahead.

She cannot see it—not yet—but she feels it. A presence, a pull, a whisper in the water that tugs at something deep within her, something she does not understand.

It is not hunger.

It is not instinct.

It is something more.

The ocean is changing. She does not know the words for warmth, for current, for movement beyond the slow, drifting pulse of the abyss. But now, there is a shift—a softness in the blackness, a thinning of the dark.

A glow.

Faint. Distant.

But there.

It is not like her own. Her lure has always been her secret, her deception, her silent promise to the blind creatures of the deep. But this is different. This light does not belong to her. It is not a trick.

It is something greater.

She does not know what she is moving toward, only that she cannot turn back.

The water feels looser now, less like the crushing embrace she has always known and more like something fleeting, something slipping through her body instead of pressing against it. The weight of the deep is fading, and with it, the world she has always understood.

And still, the light calls.

She does not know what awaits.

She does not know what it means.

But she rises toward it, all the same.

# THE WATER GROWS WARMER, THE PRESSURE CHANGES, HER BODY STRAINS—BUT SHE CONTINUES

The ocean is shifting.

It is subtle at first, a change so gradual that she does not yet understand what is happening. But her body knows.

The cold that has always wrapped around her, steady and eternal, is unraveling. The water grows softer, lighter—warmer.

She does not know warmth. She was born in the deep, where the temperature never wavers, where cold is not a sensation but a fact of existence. There, the water is unmoving, untouched by the chaos of the world above.

But now, as she ascends, the deep loosens its grip.

She feels it in her skin, in the delicate filaments of her fins. The weight that has cradled her for so long, the pressure that shaped her, **is fading.**

It is not relief. It is something else.

Something unnatural.

The walls of her body, once built to withstand the crushing depths, now sense a new kind of danger. Her tissues, perfectly balanced for a world where the ocean's embrace is inescapable, are beginning to **unravel.**

The pressure is not enough.

Her body strains, shifting in ways it was never meant to. Tiny bubbles form within her bloodstream, air expanding where there should be none. The very balance of her being—the delicate equilibrium that has kept her alive—**is breaking.**

But she does not stop.

She does not know that she is dying.

She does not know that what calls her is beyond her reach, beyond what she was made for.

She only knows that something is pulling her forward.

And so, even as the water around her grows warm,

even as the invisible weight that has always held her together begins to slip away —

**She continues.**

# THE COST OF THE ASCENT—WHY THE DEEP CANNOT BE LEFT BEHIND

The deep sea does not forgive those who try to leave it.

It is a world of immense pressure, where every creature is sculpted by the weight of the ocean pressing down upon them. For those who belong to the abyss, survival is not just about hunting or hiding—it is about **existing in equilibrium** with the crushing forces that define their world.

But when that balance is broken, when a creature built for darkness begins to rise toward the unknown, the deep does not follow.

It lets go.

And in doing so, it unravels everything it once held together.

## The Silent Killer: Pressure and the Fragility of the Deep

For a creature of the abyss, pressure is not a burden—it is a part of them, woven into their very biology. Their flesh is soft, almost gelatinous, adapted to exist in a world where external force shapes every movement, every breath, every moment. Their bones, where they exist at all, are not built for resistance but for yielding, for bending to the silent, inescapable hand of the ocean's weight.

But in the shallower waters above, that hand releases.

The body, no longer compressed by thousands of pounds per square inch, begins to change. It swells. Expands. Tissues designed to withstand immense external force now push outward, **rupturing, tearing, distorting.** Organs shift unnaturally, their delicate structures not meant for freedom, but for confinement.

And inside the bloodstream, a far greater danger begins.

Bubbles.

Tiny, imperceptible at first, forming deep within veins, within cells—air that should never be there. The same force that bends steel, that crushes submarines, is suddenly absent. Without it, gases expand, pushing against the fragile walls of arteries, pressing into places they were never meant to go.

This is **barotrauma**, the silent assassin of deep-sea creatures who rise too quickly, the unseen force that **rips them apart from the inside.**

She does not know this.

She does not understand the breaking, the slow unraveling of her form.

She only knows she is rising.

## Heat, Light, and the Stranger's World

With each slow pulse of her fins, the water grows warmer. It is not the warmth of life, not the steady pulse of a body keeping itself alive—it is something foreign, something invasive.

In the abyss, heat is a rare and violent thing. It comes only from hydrothermal vents, fissures in the Earth that bleed molten energy into the void. And there, only the toughest creatures survive—those that have made a pact with the fire, those that feed on minerals instead of flesh.

But the warmth she feels now is different.

It is not the deep's fire. It is something from above.

She does not know that warmth comes with consequences. That for a creature made of liquid and shadow, temperature is as dangerous as pressure. Heat weakens, heat dissolves. The delicate balance of proteins, of membranes, of life itself—**all begin to fail.**

And then, for the first time, something glows ahead.

Not the soft flicker of her own lure.

Not the false promises of another predator.

Something vast. Something she does not understand.

**The surface.**

She is close now.

Too close.

And still… she keeps rising.

## AND YET... SHE KEEPS RISING

She is breaking.

She does not know it, but the ocean does.

The deep that once held her together, that wrapped around her like an unspoken promise, **is gone.** The pressure that shaped her, the weight that made her whole, **has abandoned her.**

And in its absence, she is unraveling.

Her body swells, the delicate balance of her form no longer tethered to the abyss. Inside her veins, tiny bubbles bloom, spreading like silent fire through the channels of her being. Her skin—once soft, pliable, designed to withstand the relentless force of the deep—begins to stretch, to strain.

And yet… she does not stop.

She does not understand what is happening. There is no word for pain in her world, no concept of suffering, no recognition of what it means to be unmade.

She only knows **the call.**

It is there, just ahead. The pull, the glow, the unseen force that beckons her forward. The light that is not hers. The warmth that is not her own.

She has never felt it before. She has never known anything beyond the silence of the depths. But now, as her body rebels, as the ocean releases its grip on her, **she is closer than ever before.**

She should not be here.

And yet…

**She keeps rising.**

The dark water around her begins to thin. Shapes form in the distance—blurry, shifting, moving in a way the abyss never has. The silence that once cradled her is no longer absolute. There is something else here, something foreign, something she cannot name.

She has never questioned her place.

She has never fought against the deep.

She has never dared to leave behind the world that

made her.

And yet…

She **has.**

She is somewhere new now. Somewhere different. Somewhere no anglerfish was ever meant to be.

And still…

**She keeps rising.**

# CHAPTER 3: FIRST TOUCH OF LIGHT

## A GLOW THAT IS NOT HER OWN

For the first time in her existence, the darkness is not absolute.

She does not understand it at first. The deep has always been a world of absence—of light, of warmth, of anything beyond the steady rhythm of hunger and waiting. But now, in the water around her, there is something else.

A glow.

It is not the familiar flicker of her own lure, the tiny beacon that has always danced before her, casting its lonely shimmer into the void. This is different. This is **everywhere**.

It is not small. It is not fleeting. It does not waver like the false promises of the deep.

It **surrounds her.**

And for the first time, she is seen.

Her skin, which has never known illumination, which was shaped only for shadow, is now touched by something new. The soft undulations of her fins, the delicate filaments that trail behind her, the curve of her body—they are all revealed, no longer mere impressions in the black, but something **real.**

She does not know the word for color.

She does not know the concept of reflection, of surfaces, of a world beyond the one she has always known.

But she knows **this**—the feeling of something different, something unnatural pressing against her, wrapping around her like an unseen current, pulling her into a world she does not understand.

It does not feel like home.

It does not feel safe.

And yet, it is beautiful.

She does not yet know that she is near the surface. That the place she has come to is one no anglerfish has ever reached before.

She only knows that for the first time in her existence, she is **bathed in light.**

## FOR THE FIRST TIME IN HER EXISTENCE, SHE IS BATHED IN LIGHT

Light does not exist in the abyss.

It is not a thing that fades, nor a thing that comes and goes. It is simply not there. It is a silence, an absence, an impossibility.

Until now.

Now, it is everywhere.

She does not have the words for what is happening. Her body has never been touched by anything but the dark, cradled by a world where sight is meaningless, where shadow is the only constant. But now—**she is exposed.**

The light does not flicker like her own. It does not pulse like the bioluminescent ghosts of the deep. It does not belong to her, and that is what makes it terrifying.

Because in this moment, for the first time, **she is no longer hidden.**

The veil of blackness that has kept her invisible, the shield of eternal night that has cloaked her from the eyes of anything greater—**it is gone.**

And she is **seen.**

She does not know the word for fear.

But she feels it now.

Her delicate skin, made for the weight of the abyss, **feels the warmth.** The pressure of the deep, the embrace that has always held her together, is gone. She is unraveling, her body shifting, expanding, stretching in ways that were never meant to happen.

The light reveals everything.

Her teeth, jagged and curved like broken glass, now gleam in a way they never have before. Her eyes— **small, useless in the dark**—are no longer blind. They drink in something they were never meant to perceive, and they do not know how to understand it.

The surface is closer now.

Too close.

She should not be here.

She should not have come this far.

But the moment she reached the light, there was no going back.

Because once you have seen the glow of the un-known, once you have **felt something greater than the world you were born into—**

**How can you ever return to the dark?**

And yet, even in the beauty of this revelation, some-thing is breaking.

She does not know it yet.

But she is dying.

And still… she rises.

# WHAT DOES IT MEAN TO STEP BEYOND WHAT YOU'VE ALWAYS KNOWN?

To leave the deep is to leave behind everything.

The silence. The certainty. The darkness that was never just a place, but a part of her, woven into her existence like breath into lungs.

She was never meant to rise.

And yet, she did.

She did not know what called her upward. She did not know why the stillness of the abyss—her home, her safety—had become a cage she could no longer bear. But something had stirred, something deeper than hunger, older than instinct. A whisper in the water. A pull.

And she had followed it.

Now, she exists between two worlds—**the world she was made for, and the one she was never supposed to reach.**

She does not know what it means to break free from the place that shaped her. She does not know that there are names for such things: **Curiosity. Discovery. Fate.**

She does not know that beyond the abyss, beyond even the ocean, there are creatures that would recognize this feeling—**the ones who walk the earth, the ones who gaze at the stars and wonder if they, too, belong to something greater than the world they were given.**

She does not know the cost.

That to step into the unknown is to be changed by it.

That to seek the light is to risk never returning to the dark.

That there is beauty in breaking free—**but there is also danger.**

And as she rises, as her body strains against the impossible, she is no longer just an anglerfish, no longer just a drifting shadow in the black.

She is something else.

Something **that dared.**

Something that touched the edge of another world.

Something that will not return.

## THE COST OF SEEKING THE UNKNOWN—THE BEAUTY AND THE DANGER OF IT

She has left behind the deep.

She has abandoned the world that shaped her, the place where she was invisible, where she was **safe.**

And now, she is here. **Exposed. Vulnerable. Breaking.**

But there is beauty in the breaking.

The light is all around her now, not just something distant, something imagined, but something she has **touched.** Something that has **touched her back.**

It is warm. It is vast. It is infinite.

And yet, it is killing her.

The ocean is a world of balance, of laws that cannot

be defied. For those who belong to the abyss, there is **no leaving unscathed.** There is no rising without cost. The pressure that once held her together is gone, and now she is unraveling—**not because she was weak, but because she was never meant to be here.**

But if this was never meant to be…

Then why does it feel so **right?**

The unknown is a paradox. It calls to those who were never supposed to answer. It tempts, it beckons, it whispers of something greater, something more.

And yet, it does not forgive.

The cost of seeking **is always sacrifice.**

To rise is to fall. To touch the light is to surrender to it. To step beyond what you were made for is to leave behind the certainty of who you were.

And perhaps that is why the world mourns her.

Why they watch her final moments with reverence, why they see her not as something lost, but as something **that dared.**

Because she is all of us.

The ones who reach for the impossible.

The ones who leave behind comfort for the chance to touch something greater.

The ones who know the price—**and still rise anyway.**

## BUT AS SHE REACHES THE SURFACE, SOMETHING HAPPENS...

The water is no longer endless.

For the first time, there is something **above** her. A limit. A threshold. A world beyond the ocean that she cannot begin to understand.

And then—**she breaks through.**

The surface **shatters around her**, rippling outward in perfect circles, as if the sea itself gasps at her arrival. For a moment, she is caught between two realms—the abyss that made her and the world that was never meant to know her.

She is weightless.

She does not know the sky. She does not know air, or wind, or the stars that watch from above. She does not understand this place where the ocean ends and something else begins.

She only knows **light.**

Blinding. Relentless. Overwhelming.

It sears into her skin, into her bones, into every fragile part of her that was never meant to feel its touch. The warmth that once seemed so distant is now an **unforgiving fire**, consuming her, burning through her delicate form in ways she cannot resist.

She cannot breathe.

The water that has always carried her, that has always surrounded her like an eternal embrace, is now beneath her, slipping away. She does not understand **emptiness.** The weightlessness that once felt freeing now feels **like falling.**

Her body is failing.

The pressure that held her together is gone. Her insides, shaped by a world where force is law, are expanding, tearing, **coming undone.**

She has touched the light.

And it is too much.

But in this moment—**this brief, impossible, beautiful moment—she exists in a way no anglerfish has before.**

Not hidden. Not unseen.

But in the open. In the glow.

And though she does not know it, the world is watching.

A pair of human eyes catches her final movement—a flash of something strange, something foreign, something impossibly rare. A camera lens blinks, and **her last moment is captured.**

She has risen beyond the abyss.

She has been seen.

And now, **she will never be forgotten.**

# CHAPTER 4: **THE FINAL MOMENTS**

## THE LIGHT IS BLINDING, WARM, CONSUMING

The deep has never hurt her.

The abyss, for all its vastness, for all its silence, has never been cruel. It has never burned, never pierced, never taken more than what was necessary. It was dark, yes. It was cold, yes. But it was home. It was **hers.**

The light is not.

The moment she breaks the surface, it is **everywhere.**

It is not soft. It is not gentle. It is **overwhelming.**

It reaches into every part of her—her skin, her eyes, her fragile body that was never meant to feel such things.

The warmth that once whispered from a distance now **devours.**

It does not cradle her. It does not hold her as the deep once did. It **consumes.**

She cannot escape it. She cannot hide. There is no shadow to slip into, no comfort in the unseen. For the first time, **there is nothing between her and the world.**

And it is too much.

Her skin begins to dry, delicate membranes meant only for the embrace of water now **exposed to something foreign, something lethal.** The very thing that makes life possible for others—**light, warmth, air**—is what is **undoing her.**

The ocean, the only thing that has ever known her, that has ever held her together, **is slipping away.**

And yet—

For one moment.

For **one single moment**—

She is **here.**

Not hidden. Not unseen. Not waiting in the dark.

She is something **that has risen.**

Something **that has touched the impossible.**

And even as she begins to break apart, even as the deep calls for her to return, she does not sink.

Not yet.

Not before she feels what no other of her kind has ever felt.

**Not before she exists in the light.**

## HER BODY IS FAILING—SHE WAS NEVER MEANT TO BE HERE

She does not know what is happening.

She does not know the word for suffering, for decay, for the slow unraveling of her own existence.

But she feels it.

Something inside her is breaking, shifting in ways that it never has before. Her body, so perfectly shaped by the abyss, so carefully balanced by the weight of the deep, **is coming undone.**

She was never meant to exist without pressure.

The ocean has always **held her together.** It has wrapped around her like an unspoken promise, pressing into her flesh, into her bones, into the very fibers of her being. It was the only reason she survived. It was the force that shaped her, that gave her form, that dictated what she could and could not be.

But now, that force is **gone.**

Her body is no longer held. No longer contained.

The water here is loose, soft, too light, too empty. The deep's relentless grip has been replaced by something gentler—but **gentle does not mean safe.**

Her cells are **expanding.**

Her tissues are **swelling.**

The same pressure that once kept her whole has abandoned her, and now **there is too much space inside her.** The very fabric of her being is **stretching, tearing, unraveling in silence.**

She does not understand pain.

But she understands this:

She cannot go back.

She cannot sink into the place she came from, cannot return to the silence, to the cold, to the blackness that was once her world. Her body is no longer what it was. The deep would not take her now.

She does not belong there anymore.

She does not belong **here** either.

She was never meant to be anywhere **but the place she has already left behind.**

And yet, even as her body betrays her, even as the very ocean that made her begins to let her go—

She does not fight.

She does not struggle.

Because in this moment, **she has already won.**

# BUT FOR A MOMENT, SHE EXPERIENCES SOMETHING NO ANGLERFISH HAS BEFORE

For a moment, she is **more than what she was made to be.**

More than a creature of the abyss. More than a drifting shadow in the dark.

For a moment, she is **part of something greater.**

The light does not belong to her, yet it **holds her now.** It wraps around her fragile body, filling the spaces where darkness once lived, pressing against her with an embrace that is both violent and beautiful.

She does not know the word for wonder.

She does not know what it means to **witness something new,** to step beyond the boundaries of her world and into something impossible.

But she **feels** it.

The warmth of it. The weightlessness of it. The way

it pours over her skin, highlighting the ridges of her bones, the delicate curve of her fins, the flickering remnants of her own fading glow.

For a moment, **she is not just an anglerfish.**

She is something **that has risen.**

Something **that has reached beyond its world.**

Something **that has seen.**

She does not know that there are others watching. That eyes above the water have caught sight of her strange and alien form. That for the first time, her existence is **not hidden.**

That for the first time, she is **witnessed.**

And maybe—**that is enough.**

Even as her body fades, even as the deep calls for her return, **she has touched the impossible.**

And no matter what happens next—**she will never be forgotten.**

# WHY DID SHE DIE? THE PHYSICAL TOLL OF DEEP-SEA CREATURES REACHING THE SURFACE

The ocean is a world of rules.

Not the kind that can be broken. Not the kind that can be rewritten.

These are the rules of pressure, of depth, of life shaped by forces older than time itself. And for those born in the abyss, the greatest rule of all is this:

**You do not leave.**

She did.

And the ocean did not follow.

It let her go.

And in letting go, it took everything from her.

## The Science of the Deep—And the Cost of Leaving It

At the depths where she was born, the water does

not simply surround—it **presses.** It is a force, a weight, a silent presence that does not relent. Thousands of pounds per square inch shape every part of her, molding her body into something that can **only** exist beneath that crushing embrace.

Her bones are not like those of creatures that swim in the shallows. They are weak, soft, more suggestion than structure. Her flesh is delicate, designed not to resist pressure, but to **exist within it.** She is not built to fight the weight of the ocean—**she is built to become a part of it.**

But the moment she left, the moment she **rose beyond where she was meant to be,** that balance collapsed.

Her body was not held together by muscle or skin alone—it was held together by the deep itself. And now, without it, she is **expanding.**

The gases in her bloodstream, once safely compressed by the weight of the abyss, have begun to **grow.** Tiny bubbles, invisible at first, stretch and spread, pushing against the fragile walls of her veins, tearing through them like unseen daggers.

This is **barotrauma**—a quiet, merciless unraveling.

Her organs swell. Her eyes, unused to light, **bulge from their sockets.** Her stomach, once compressed within the tight embrace of pressure, **turns outward.**

She is breaking.

Not because she is weak.

But because **she was never meant to be free.**

And yet, even in this unraveling, even in the silent devastation of her body turning against itself, there is no struggle.

She has already risen.

She has already touched the light.

And **maybe that was always the cost.**

## BUT MAYBE... THIS WAS MEANT TO HAPPEN

The ocean does not make mistakes.

It is vast, unknowable, endless—but it is not careless. It does not bend to impulse, does not leave room for those who stray beyond their place.

And yet, she did.

She rose when no others had risen.

She saw what no others had seen.

And now, even as she drifts between two worlds, between the deep that made her and the light that unmade her, she is **still.**

There is no fear in her. No fight. No struggle.

Only **existence.**

Only the quiet certainty that she has become something else.

Something more.

She was not meant to be here.

And yet… she is.

And maybe, just maybe—**this was always how it was meant to be.**

Maybe she was never meant to die unseen, swallowed by the dark like the millions before her.

Maybe she was meant to **rise, even if only once.**

To leave the abyss and reach beyond what she was told she could be.

To be caught in the eyes of something greater, something watching, something that would remember.

Because she is remembered now.

She will never return to the depths, never sink into the silence, never fade into the forgotten.

She has touched the light.

And the world will never forget that she did.

# CHAPTER 5: **THE LEGEND BEGINS**

## A DIVER FILMS HER LAST MOMENTS

The ocean is quiet here.

Not like the abyss, where silence is a presence, a weight pressing down from all sides. Here, it is a different kind of quiet—the kind that moves, that shifts with the currents, that carries the echoes of life in a world where the sun still touches the waves.

And within that gentle quiet, she drifts.

A shadow in the shallows. A visitor from a place no one was ever meant to see.

She does not know that she is being watched.

Not by the predators of the deep, not by the ghostly hunters of the abyss.

But by something **else.**

Something **that does not belong to the ocean.**

A shape moves above her—slow, deliberate. A figure clad in artificial skin, suspended in the water like a creature that does not know if it belongs to the sea or the sky. A human.

The diver does not yet understand what they are seeing.

They know the ocean. They have seen its wonders, its secrets. But this?

This is different.

They know, instinctively, that this is something **rare.**

They lift the camera.

A mechanical blink. A lens focuses.

And in that instant, **her last moment is captured.**

The anglerfish drifts, her body floating just beneath the surface, her once-powerful form now fragile, unmade by the very world she dared to reach.

Her lure, dimming now, sways gently in the current.

Her fins, delicate as whispers, move no more.

She is still.

But she is **seen.**

And with the press of a button, her story begins.

## THE INTERNET SEES HER. THE WORLD WEEPS.

The ocean forgets.

The waves do not pause to mark her passing. The currents do not whisper her name. The deep does not grieve.

But **above the water, far beyond the reach of the sea, the world watches.**

The diver's video is uploaded—just a fragment of time, a flicker of movement, a brief glimpse of something that should have never been there.

At first, it is just another clip. Another moment lost in the endless tide of images, sounds, distractions. A creature, alien in its beauty, floating in the light.

But something about her lingers.

Something about **her story** refuses to fade.

The video spreads—passed from hand to hand, screen to screen, until it is **everywhere.**

People watch. They stare into the glassy black of her eyes, at the curve of her fragile fins, at the way her body drifts, weightless, unmoored.

And something **unexpected happens.**

They feel.

A sadness they do not fully understand. An ache, an emptiness, a quiet mourning for a creature they have never met.

She is not fierce. She is not mighty. She is not a predator whose final moments inspire fear or awe.

She is something else.

She is **delicate.**

She is **small.**

She is **out of place.**

And in her, they see something of themselves.

They see **struggle.**

They see **reaching for something beyond your world.**

They see **what it means to step beyond the familiar, to seek something greater—even if it means losing yourself along the way.**

The comments flood in.

*"Why does this make me so sad?"*
*"She just wanted to see the light."*
*"She didn't belong here, but for a moment... she did."*

They grieve for her.

Not because she was meant to live forever. Not because they could have saved her.

But because **they saw her.**

And now, she **will never be forgotten.**

# A DEEP-SEA CREATURE THAT LIVED AND DIED IN DARKNESS— NOW IMMORTALIZED IN LIGHT

She was never meant to be seen.

The deep does not grant its secrets freely. It hides them beneath miles of water, in the silence, in the pressure, in the dark that no sunlight can pierce. For millions of years, she existed where no eye could follow, where no witness could mourn her passing.

And yet—**here she is.**

Suspended in the glow.

A creature of the abyss, meant only for shadow, now held forever in the radiance of human memory.

She did not rise to be known. She did not seek an audience, did not hunger for attention, did not rise from the deep for anything more than an instinct she could not name.

But now, **she is everywhere.**

Her image is shared in waves, a ripple spreading

across the world, **farther than she ever traveled, farther than she ever could.**

In the depths, she was just another phantom, another flicker in the black, lost to the cycle of survival.

But here, in the light—**she is a legend.**

Her body, so fragile, so fleeting, may be gone. But the moment she broke the surface, the moment the camera captured her final drift, she became something more than herself.

She became a story.

And stories do not sink.

They do not disappear beneath the waves. They do not vanish into the abyss.

They live.

She **lives.**

Not in the ocean. Not in the deep.

But in every heart that saw her.

In every tear shed for a creature who never knew

she was watched.

In every whisper of awe at the sight of something that should never have been here—**but was.**

She lived in darkness.

But now, she belongs to the light.

Forever.

## HER JOURNEY BECOMES A METAPHOR FOR STRUGGLE, SACRIFICE, AND SEEKING SOMETHING GREATER

The ocean is full of stories.

Tales of survival, of predators and prey, of endless cycles that repeat without witness, without memory, without meaning beyond the hunger that drives them.

But **hers is different.**

Because **she rose.**

Not to hunt. Not to escape. Not even to survive.

She rose for **something unknown.**

And in that, she became more than just a creature of the deep.

She became **a symbol.**

A symbol of those who leave behind everything they know in search of something greater.

A symbol of those who reach beyond their world, even when they do not understand what waits for them.

A symbol of those who rise—**even at great cost.**

People watch her final moments and see something beyond biology, beyond science. They see **themselves.**

The dreamers. The wanderers. The ones who crave the impossible, even when they know it might break them.

She is every soul that has ever **left the familiar behind, chasing a light they cannot name.**

She is the artist who creates without knowing if

anyone will understand.

She is the explorer who walks beyond the map's edge, knowing there is no way back.

She is the child who reaches for something just beyond their grasp, not knowing if it will hold or if they will fall.

She is **the cost of seeking.**

She is **the price of wonder.**

She is **the proof that some things must be seen, must be felt, must be touched — even if they cannot be kept.**

And though she is gone, her story is **not.**

Because stories do not drown.

Stories **rise.**

And so did she.

# WHY ARE WE SO CONNECTED TO CREATURES LIKE THIS? WHAT DOES THIS SAY ABOUT US?

It was just a fish.

A creature born in the cold, in the dark, in a place unseen by the world above.

And yet, **we grieved for her.**

Why?

Why did a single anglerfish—a species we have rarely thought about, a being so far removed from our world—**stir something deep within us?**

Perhaps it is because we recognize her journey.

Because in her rise, we see our own.

## The Science of Connection—Why We Mourn the Unknown

Human beings are wired for stories. Our brains are not just processors of logic and survival—we are creatures of meaning. We do not simply observe;

**we interpret.**

We see a small, fragile fish break the surface, and we do not just see an anglerfish.

We see **a soul that reached beyond itself.**

This is what **biologists call biophilia**—our innate emotional connection to other living things. It is why we cry for creatures we will never meet, why we mourn a whale lost at sea, why we ache at the sight of a dying bird, even when we do not understand its song.

We are drawn to stories of struggle, of reaching, of rising **against what is natural, what is expected, what is safe.**

And she did that.

Her journey was not **human,** but it was **human-like.**

She was small. She was vulnerable. She was insignificant in the vastness of the world.

And yet—**she dared.**

And we saw her.

For a species that has always looked beyond itself, that has always wondered what lies **past the horizon, past the sky, past the limits we were given—**

How could we not see ourselves in her?

She is the astronaut drifting beyond Earth's atmosphere, staring into the black unknown.

She is the explorer who steps beyond the edge of the map, into lands unseen.

She is the artist who creates, not knowing if the world will ever understand.

She is every moment of reaching.

And she is the reminder that sometimes, **there is no going back.**

Not because we failed.

Not because we were lost.

But because the act of rising **changes us forever.**

She is gone.

But she is **not forgotten.**

Because **she rose.**

And that is all that ever mattered.

# EPILOGUE: **THE FISH WHO TOUCHED THE LIGHT**

The deep has not changed.

It is still there, stretching endlessly beneath the waves. The currents still drift unseen, the silence still hums in the darkness, and life continues as it always has—hidden, untouched, unremembered.

But **she is no longer part of it.**

She did not return. She did not fade into the abyss, swallowed once more by the black.

She rose.

And the moment she did, she became something **more than the deep, more than the light, more than just another fleeting life in the vast and endless ocean.**

She became **a story.**

Her journey was never meant to be seen. And yet, we saw it.

And now, we carry it.

Because hers was not just the story of a fish that broke the surface.

It was the story of **what it means to seek, to reach, to move toward something you do not yet understand.**

It was the story of **the ones who dare.**

The ones who leave behind what is safe for what is unknown.

The ones who cannot explain why they must go, only that something calls them forward.

The ones who rise.

Some will say she was lost.

That she should never have left the deep. That she should never have touched the light.

But maybe, just maybe—**she wasn't lost at all.**

Maybe she found something greater.

And maybe, in watching her, in mourning her, in remembering her—**so did we.**

# ACKNOWLEDGMENTS

To **her.**

To the anglerfish who rose, who touched the light, who showed us what it means to seek the unknown. You did not know the world was watching, but now, **it will never forget.**

To the diver—the one who captured her final moment, who unknowingly turned a fleeting life into a story that will live forever.

To the ocean—endless, mysterious, and filled with stories waiting to be told.

To the ones who saw her—who felt something stir within them as she drifted toward the light.

To the dreamers, the seekers, and the wanderers— may you always chase what calls to you, even if you do not yet understand why.

And to **you, dear reader.**

You are part of this story now.

Because stories do not sink.

**They rise.**

With gratitude,
**Felix Grayson**

# ABOUT THE AUTHOR

Felix Grayson is a storyteller fascinated by the moments that captivate the world. From the strange and unexpected to the deeply moving, he explores **the stories that capture our collective imagination — those viral events that spread across the internet and leave us thinking, wondering, and feeling.**

With a passion for uncovering meaning in the seemingly random, he transforms fleeting headlines into **thought-provoking, immersive narratives** that stay with readers long after the moment has passed. His work blends **science, emotion, and storytelling** to bring depth to the viral stories we can't stop talking about.

When a moment **takes the world by storm,** he's there — ready to tell the story behind the story.

www.ingramcontent.com/pod-product-compliance
Lightning Source LLC
Chambersburg PA
CBHW032103020426
42335CB00011B/464